Luiz Antonio Fraga Moreira
Marcelo Massarani

Reliability Centred Maintenance applied to Vehicles

Luiz Antonio Fraga Moreira
Marcelo Massarani

Reliability Centred Maintenance applied to Vehicles

A practical application of Reliability Centred Maintenance

ScienciaScripts

Imprint

Cover image: www.ingimage.com

This book is a translation from the original published under ISBN 978-3-330-76721-8.

Publisher:
Sciencia Scripts
is a trademark of
Dodo Books Indian Ocean Ltd. and OmniScriptum S.R.L publishing group

120 High Road, East Finchley, London, N2 9ED, United Kingdom
Str. Armeneasca 28/1, office 1, Chisinau MD-2012, Republic of Moldova, Europe
Managing Directors: Ieva Konstantinova, Victoria Ursu
info@omniscriptum.com

Printed at: see last page
ISBN: 978-620-8-56677-7

I dedicate this study to my family, to the team at The Specialist and to the Land Rover vehicle users who collaborated on this work

ACKNOWLEDGEMENTS

To Prof Dr Marcelo Massarani, for his patience and companionship during the preparation of this study, to my wife and children, for putting up with the long work, and to Cristiane, for her dedication and commitment.

"Success consists of going from defeat to defeat without losing enthusiasm."

(Winston Churchill)

SUMMARY

A number of studies have been carried out into the implementation of Reliability Centred Maintenance (RCC) in complex systems, either to increase the safety of the people involved and the environment around the installations, or to reduce maintenance costs. These studies mainly address its implementation in civil, military and space aviation aircraft, as well as in power generation companies, and particularly nuclear power plants. This work was conducted to implement the MCC system in the automotive field, more specifically in monitored maintenance fleets and vehicles travelling in inhospitable locations that are difficult to access for rescue in the event of a failure. A statistical survey determined the useful life of the water pump in the case study and recommended its inspection. The conclusion is that it is possible to implement CCM on certain components by carrying out a life analysis and specifying the correct time for inspection and replacement of the component before it fails - which can be critical and cause damage to other parts of the vehicles.

Keywords: reliability centred maintenance (applications), vehicles.

SUMMARY

CHAPTER 1

INTRODUCTION

Maintenance techniques evolved little during the early years of the industrial revolution (SMITH, 1993). Equipment was only repaired when it stopped performing its function. Due to their small workload, the operators' extensive knowledge of the equipment they were operating and the excesses characteristic of the time - oversized projects - there was no need for a specific maintenance team and this type of maintenance met the necessary standards.

Commonly referred to as Corrective or Reactive Maintenance (NASA, 2000), this type of maintenance can still be adopted today, but it is only recommended that it be carried out on components that are not liable to completely paralyse the system or that do not compromise the performance of other components or the safety of operators or the environment. In this system, a technical and economic-financial analysis is carried out to assess whether the cost of a shutdown, for example, is worthwhile compared to other types of maintenance.

Soon after the Second World War, most companies began to implement another maintenance system for their equipment, respecting the manufacturers' instructions, which determined the ideal time for replacement before failure, saving resources and reducing downtime for repairs (SMITH, 1993).

The equipment of the time, most of whose components and mechanical devices were simple, had two characteristic behaviours: the first with a high probability of failure at the start of activity, a constant and low probability of failure throughout most of its useful life, and a considerable increase in the probability of failure towards the end of its useful life, when wear and tear increased (Graph 1 FDP type E - NASA, 2000); the second behaviour implied a low probability of failure at the start and a high probability at the end of its life (Graph 1).

Once the time at which the inflection in the curve occurred (increased probability of failure) was determined and the component was replaced at a time prior to the failure, the failure was avoided and, as this task could be programmed, equipment downtime took less time, allowing the maintenance team to plan the necessary resources and spare parts, generating considerable savings.

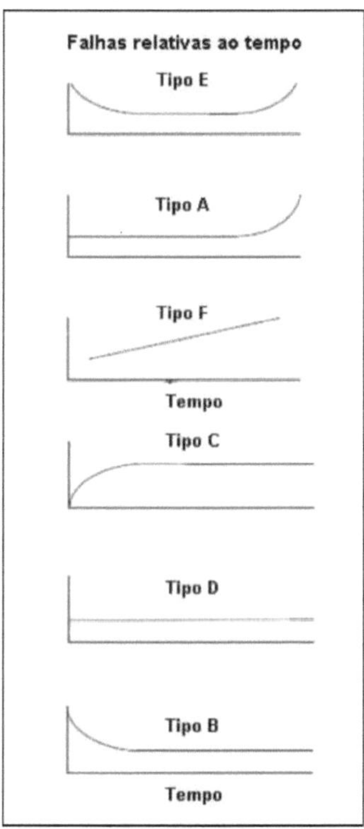

Graph 1 - Failure probability density curves over time

Source: Adapted from NASA, 2000

This type of maintenance, known as Preventive Maintenance (NASA, 2000), is usually restricted to certain parts of the equipment or system, as only components that can have a known useful life can be maintained in this way.
maintained. The disadvantage of this system is that, in some applications, components can be replaced prematurely, mainly due to the type of use of the equipment. "In fact, it has been found that, in many cases, the life of the components has far exceeded the life realised or stated in the design" (NASA, 2000, p. 1-2).

With the improvement of equipment designs, the increased use of electronic devices and more complex interconnected systems at the end of the 1980s - which could have varied failure profiles, such as types B, C, D and F illustrated in Graph 1 - coupled with increased workloads and higher operator turnover, preventive maintenance lost its efficiency, causing breakdowns and unwanted downtime (SMITH, 1993).

A new maintenance system called Predictive Maintenance (NASA, 2000) was then developed, allowing the end of a component's useful life to be indicated by means of measuring equipment installed in the component. The disadvantage of this type of maintenance is that only components that can have their useful life monitored by measurable parameters are candidates for it, restricting its practical application.

A fourth type of maintenance called Proactive Maintenance (NASA, 2000), which uses the statistical techniques AE (useful life analysis), RCFM (root cause failure analysis) and FMEA (failure modes and effects analysis), based on the equipment and its components at the installed site and the workload of each system, has proved efficient in maintaining the most sophisticated and complex equipment in existence today.

With the growth of the aeronautics industry, the advent of civil aviation and the development of the electricity generation industry (especially nuclear power), which required optimised designs (cost improvements and weight reduction) and extremely high reliability, new levels of maintenance had to be reached, challenging the skills of technicians (NOWLAN ETAL, 1978).

According to the same authors, in order to make aircraft maintenance economically viable without jeopardising their reliability, *United Airlines* and *Boeing Commercial Airplanes* jointly developed a new technique called Reliability Centred Maintenance, the subject of this study.

The automotive industry has evolved a great deal thanks to new vehicle design techniques, the development of more efficient systems, field tests and new testing technologies (computer simulation). With the consequent correction of various problems, there has been an increase in reliability in all systems, such as cooling systems, on-board electronics, brakes and steering, especially since the early 1990s.

However, vehicles that are used intensively on the road and in adverse conditions, such as those used by fleet operators, can suffer defects that do not affect other vehicles (little travelled in relation to their lifespan), not even in the most complex simulations or field tests. Information from the field (user complaints) does not always reach designers and manufacturers in time to correct a faulty system, either due to a lack of communication and contact between user and manufacturer, or due to a lack of interest on the part of car manufacturers in establishing this contact and correcting reported faults.

This fact was exacerbated when, in the mid-1990s, the market opened its doors to imports of vehicles that circulated in their countries of origin under much better conditions and closer to the design and testing conditions than those that existed and still exist in Brazil (especially with regard to working temperatures and road conditions). Evidence of this fact is that in the state of São Paulo, for example, only around 15% of roads are asphalted (DEPARTAMENTO DE ESTRADAS DE RODAGEM, 2008).

The maintenance of these vehicles is therefore jeopardised, since the actual conditions in which the vehicles actually travel were not usually taken into account in the design.

This study is divided into five parts. The first chapter describes maintenance with its traditional systems. The second presents the MCC system (reliability centred maintenance) as initially implemented in the aeronautics and power generation industries, with its variations, advantages and disadvantages, and the process of implementing it, step by step, detailing the tools used (FMEA, block diagram, reliability, AE and LTA - logical analysis tree).

The third chapter discusses the implementation of the MCC system for road vehicle use in captive fleets that require monitored maintenance, using the same tools mentioned above.

The fourth chapter presents an example of a practical application, with data taken from repairs carried out at a company specialising in vehicle maintenance. A statistical survey of the lifespan of the component that is the focus of this study was carried out based on this data. It also discusses the validity of this application and its future use.

The fifth chapter draws conclusions. Finally, the appendix illustrates the typing of six pages, out of a total of 124, relating to 4950 components exchanged in 2164 passages at the example case study company between 2004 and 2008, from which the statistical survey data was compiled.

The general objective is to study the MCC system in vehicular applications, using the necessary tools in order to increase the reliability of vehicles, especially those used in fleets with monitored maintenance.

CHAPTER 2

RELIABILITY CENTRED MAINTENANCE (MCC)

According to Moubray (1977, p.7), Reliability Centred Maintenance (RCM) is "a process used to determine what must be done to ensure that any physical asset continues to do what its users want it to do in its current operating context". Seixas (1999, p.2) defined it as "the application of a structured method to establish the best maintenance strategy for a given system or piece of equipment".

MCC (or *RCM - Reliability Centred Maintenance)* was developed in the aerospace industry 30 years ago, and was later implemented in the electricity generation industry, mainly in plants using nuclear energy sources, due to the high degree of reliability and safety required.

The development of the MCC was based mainly on the history of previous failures in the aeronautical industry. By studying documented histories, United Airlines proved that the classic "bathtub curve", or E-curve (Graph 1), was not accurate enough to represent the failures of non-structural components of its aircraft. It was then determined that only 11% of all these components had the characteristic of probability of failure relative to time of use, while 89% failed for reasons other than fatigue (SMITH, 1993).

Based on this observation, the percentage of components replaced on time was reduced from 58 per cent in 1964 to 9 per cent in 1987, while the replacement of monitored components grew from 2 per cent in 1964 to 51 per cent in 1987, enabling the commercial aircraft industry to reduce maintenance costs and keep them low until the end of 1980 (SMITH, 1993).

The US *Federal Aviation Administration (FAA)* has accepted the new form of aircraft maintenance, which has been used on practically all new or refurbished aircraft in order to obtain a licence to fly (SMITH, 1993).

In 1977, the US Department of Defence called the new method *Reliability-Centered-Maintenance* and advised its adoption in most military systems. In mid-1980, the *Electric Power Research Institute, the* organisation responsible for assessing the maintenance of electricity generating and distribution companies in the United States, began implementing a pilot study of CCM in nuclear-powered electricity generation plants (MATTESON, 1995).

Since then, the MCC process has evolved, and the so-called "classic MCC" (commonly used by the aerospace industry) coexists with "fast MCC", the main difference between which is to reduce the amount of time it takes to acquire data (necessary for gathering a history of failures), using mainly the experience of the professionals involved in maintenance and identifying failure modes and their known causes (with studies based on *RCFA* - root cause failure analysis, *LTA* - logical analysis tree, *AE* - life analysis and *FMEA* - failure mode and effects analysis of the system and its components).

A rapid implementation of CCM can evolve, if necessary, through modifications to the process by means of a formalised and planned feedback system called a Dynamic Maintenance Programme.

The process of implementing MCC allows the maintenance manager to focus his team's main tasks on the most critical faults, saving financial and personnel resources on maintenance operations that would actually paralyse the system/equipment.

The tasks listed by the MCC study are indeed efficient: relying solely on the tasks stipulated by the manufacturer of the equipment or system is not so efficient, especially given the differences in use, environment and external conditions, which are not always tested by the manufacturer.

According to Schwan (1999), MCC aims to create strategic maintenance routines that preserve the important functions of the equipment/system in the most economical way possible. Or, "MCC is the process used to determine the most effective way to maintain" (NASA, 2000, p. 1-1).

The MCC system then determines, through its implementation, which is the best technique to be employed in the maintenance of the component or equipment (Figure 1) - whether corrective, preventive, predictive or proactive maintenance - both from a financial point of view and from a safety point of view, for the people and the environment involved (NASA, 2000).

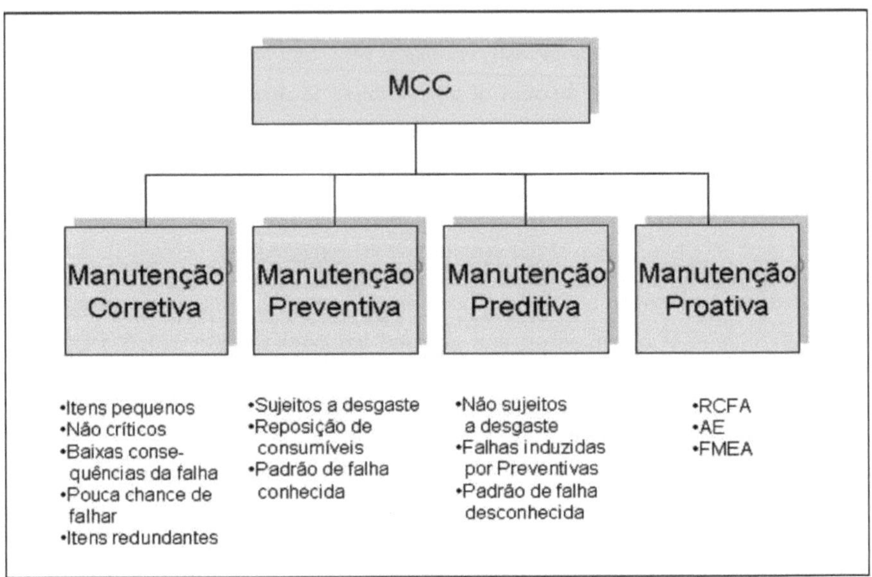

Figure 1 - Different types of maintenance and their main characteristics
Source: Adapted from NASA, 2000

2.1 IMPLEMENTATION OF THE MCC

The rapid MCC programme consists of analysing the components and equipment as described in the following seven steps:

1) *Establish the scope of the study.* Only if you know exactly which component or system will be analysed will you be able to apply the programme. When a preventive maintenance task is listed at the end of the study for a particular component, if it is not well specified, the programme will certainly fail. It is also during this phase that most of the information is collected in order to inform the following stages. According to Smith and Hinchliffe (2004), the 80/20 rule can be used to choose the scope of the study. Basically, 80 per cent of an observed effect tends to reside in 20 per cent of the available source:

> [...] .80 per cent of carpet wear is found in 20 per cent of the available carpet area - because that's where the traffic occurs. So, in a factory, 80% of reactive maintenance and production cost losses tend to be located in 20% of the systems in that factory - the so-called bad actors (SMITH, HINCHLIFFE, 2004, p. 392-413).

2) *Identify the interfaces.* Smith and Hinchliffe (2004) explain that in order to determine the influence of one component on the others, the interfaces between the components must be precisely defined in order to decide what will or will not be included in

the study. According to the authors, "[...] a clear definition of the interfaces of inputs and outputs is a necessary condition for ensuring precision in the process of analysing the system [...]" (p. 1839-1855/ This stage will provide the necessary information for the next steps, especially when using the 80/20 rule explained above.

3) *Specifying important functions.* At this stage, armed with the information previously collected, the analysts begin to separate the most critical components for maintenance, thus subsidising the manufacture of the Functional Block Diagram with complete knowledge of the inputs and outputs.

This diagram provides information on how the system is put together, and may also contain information on the specific function of each component, so as to make it clear to the analyst what could cause a problem when a fault starts. It is only a qualitative analysis, not a quantitative one.

As CCM is based on the function of each component in the system, the data generated by the information flowchart and its hierarchy become important, because only with the flowchart can the analyst know the influence of a component failure on the entire system.

According to Smith and Hinchliffe (2004), this stage comprises a complete description of the system, the execution of a functional block diagram describing the main functions of the components listed, their input and output interfaces, a complete list of the components that make up the system and their maintenance history. According to these authors, only components or systems with important functions should be studied.

4) *Identify the dominant failure modes.* Only dominant failures should be analysed. As the main function of MCC is to preserve the fundamental functions of the system, components that do not cause system downtime will not be analysed. At this stage, the focus is on the loss of function and not the loss of equipment (SMITH, HINCHLIFFE, 2004), which was the concern in the other ways of maintaining equipment.

These authors also explain that a functional failure is usually caused by more than a single loss of function, a fact that will be analysed in the next step.

Lists of components, their functions and the history of their main faults and modes are created in order to subsidise the manufacture of the FMEA in the following stages; forgetting a component at this stage will remove it from any preventive maintenance that may be listed for it (SMITH, HINCHLIFFE, 2004).

5) *Identify critical failure modes.* At this stage, once the dominant failure modes are known, the critical failures are defined. An FMEA of the equipment or component is usually carried out. This technique will allow the analyst to find out which faults will cause the equipment to stop working (loss of function). The analyst must keep in mind that the primary function of MCC is to preserve the functions of the equipment.

The following describes the stages of the FMEA process and how it should be carried out in order to apply MCC.

Once the information from the previous stages has been collected, the FMEA becomes a powerful, objective and systematic tool for analysing failure modes and causes, enabling analysts to define maintenance priorities. "The aim of an FMEA is to discover all the ways in which a product or process can fail" (MCDERMOTT; MIKULAK; BEAUREGARD, 2009).

FMEA is an important tool for preventing faults in a project, product or process before it is developed, and it is also used as a source of information for existing problems. Through FMEA, it is possible to decide which component or device should be a priority when implementing CCM, concentrating efforts and making it easier to analyse, always bearing in mind that CCM should only be implemented on the most important items and those that cause the most problems when using the equipment (CAPALDO; GUERRERO; ROZENFELD, 1999).

According to the same authors, FMEA analysis can be applied in the following situations:

- to reduce the likelihood of failures in new product or process projects
- to reduce the likelihood of potential failures (i.e. failures that have not yet occurred) in products/processes already in operation,
- to increase the reliability of products or processes already in operation by analysing failures that have already occurred, and
- to reduce the risk of errors and increase the quality of administrative procedures.

What interests this study is the third item - using FMEA analysis to increase the reliability of products or processes already in operation by analysing failures that have already occurred.

The FMEA must be conducted for each system or component identified in the definition of the interfaces, as described above. Analysing a component outside the system can be erroneous, since the context of that component's participation in one system can be completely different from the same component's participation in another system.

To illustrate this, the failure of a bearing in a particular piece of equipment may be minor and not jeopardise its main function, while in another piece of equipment the same bearing could cause a catastrophic failure (PRIDE, 2008).

Specific standards and systems guide analysts in preparing the FMEA. A criticism to be made of an FMEA drawn up for a maintenance study of a particular component is that, if it is defined that the failure mode found refers to its design, this can only be changed via the project, and therefore the analyst's access to the modification may not exist.

The process of drawing up the FMEA has different phases (CAPALDO; GUERRERO;ROZENFELD, 2005-6):

- Determine the product's functions and characteristics
- Determine the type of potential failure for each function
- Determine the effect of the type of fault
- Determine the possible cause of the fault
- Determine the current controls to mitigate the fault
- Assess the risks of the fault and classify the Severity (S), Occurrence (O) and Detection (D) indices, according to Tables 1, 2 and 3.
- Calculate the risk index by multiplying the three indices determined above with each other
- Propose improvement measures
- Prepare and display this data on an appropriate form.

In the case of an FMEA focused on products and, more specifically, on a component of a system or equipment under maintenance, the analyst can propose new indices and not stick to tables such as those suggested below (CAPALDO; GUERRERO; ROSENFELD, 2005-6) because, as explained above, the same component applied to another system/equipment can, when it fails, have a greater severity in one piece of equipment than in another (PRIDE, 2008).This study used the severity, occurrence and detection indices recommended by the above authors, illustrated in Tables 1, 2 and 3 below,

Table 1 - Severity Classification

index	*Severity*	*Criteria*
1	Minimum	*The* user barely realises that the fault has occurred.
2, 3	Small	Slight deterioration in performance with slight user dissatisfaction.
4, 5 or 6	Moderate	Significant deterioration in the performance of a system, with user dissatisfaction.
7, 8	High	System stops working and great user dissatisfaction.
9, 10	Very high	Same as above, but affecting security.

Source: Adapted from Capaldo, Guerrero and Rozenfeld (2005-6)

Table 2 - Occurrence Classification

index	*Occurrence*	*Proportion*	*CpK*
1	Remote	1:1.000.000	> 1.67
2	Small	1:20.000	> 1,00
3		1:4.000	
4	Moderate	1:1.000	< 1,00
5		1:400	
6		1:80	
7	High	1:40	
8		1:20	
9	Very high	1:8	
10		1:2	

Source: Adapted from Capaldo, Guerrero and Rozenfeld (2005-6)

Table 3 - Detection Classification

index	*Detection*	*Criteria*
1; 2	Very large	It will certainly be detected
3; 4	Big	High probability of detection
5; 6	Moderate	It will probably be detected
7; 8	Small	It probably won't be detected
9; 10	Very small	It certainly won't be detected

Source: Adapted from Capaldo, Guerrero and Rozenfeld (2005-6)

6) *Identify the dominant causes of failure modes.* Only the dominant causes of the failure mode will be studied; those that do not cause loss of function will not be analysed. In this phase, according to Smith and Hinchliffe (2004), carrying out an LTA (Figure 2) helps in the decision by classifying the failure in relation to what it can cause to the plant or system.

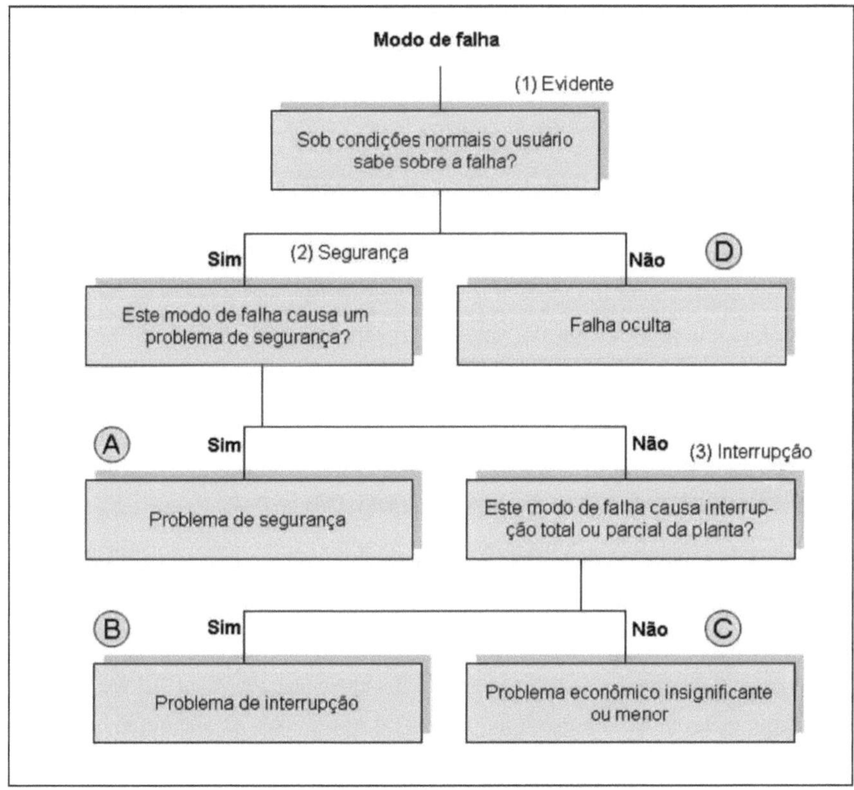

Figure 2 - LTA Fault analysis tree structure
Source: Adapted from Smith and Hinchliffe, 2004

According to the structure illustrated above, the analysis is carried out according to the failure mode and what it causes to the component and the system or equipment.

All this analysis, according to Smith and Hinchliffe (2004), is carried out for each failure mode and the result is the classification of this type of failure into three categories: the first relating to safety failure, the second to failure to interrupt the operation of the plant (or equipment) and the third relating to financial problems .

Each failure mode is entered in the first box with the question "Under normal conditions, does the user know about the failure?". If the answer is no, this failure mode is classified as "D" (hidden) and will be subject to preventive maintenance or preventive inspection, or even proactive maintenance.

The same failure mode (hidden or not, according to the previous classification) goes on to the second question, relating to safety: "Does this failure mode cause a safety problem?" If so, the failure mode is classified as "A". If so, the failure mode is classified as "A". This refers to safety in the sense of causing injury or risk to the user or surrounding personnel.

The same failure mode classified in the previous process for safety (regardless of the outcome) is then classified for economic problems. If the failure mode causes an interruption in the operation of the plant, it is classified as "B", and if not, it is classified as "C". The economic analysis of what the value of a plant interruption problem would be is relative, depending on the parameters of each company and must be analysed within its context.

The analyst classifies the failure modes according to the letters they have been assigned in the analysis (they can be types A; B; C; D/A; D/B or D/C).

According to Smith and Hinchliffe (2004), this information will be used to classify the most important failure modes that cause the most damage. For example, failure types classified as "C" should be assigned for repair only after the failure (corrective maintenance).

According to this analysis, only failure modes classified as "A" or "B" will move on to the next stage, since those classified as "C" will be designated as subject to preventive maintenance, and those classified as "D" have already been classified as preventive maintenance, preventive inspection or proactive maintenance.

7) *Select maintenance tasks.* In this phase, it is up to the analyst to define which maintenance system will be adopted for the components or system being analysed. If no type of maintenance - between Preventive, Predictive or Proactive - is chosen, the only remaining option is Corrective maintenance (SMITH; HINCHLIFFE, 2004) or RTF - operate until failure, this analysis being carried out at all levels (safety, plant shutdown or economic condition) as explained in the previous section.

To properly implement CCM, it is up to the analyst to carry out each of the steps described above, using the appropriate method to reach the necessary conclusion that will lead to the best way to maintain the component or equipment being analysed, always bearing in mind the preservation of the equipment's functions and not the equipment itself.

The maintenance task to be carried out at this stage should be:

1) *Applicable:* The task will prevent or mitigate the fault by detecting or discovering a hidden fault.

2) *Effective:* The task will be as economically efficient as possible compared to other tasks.

If the FMEA or LTA provide the analyst with sufficient information (along with the previous steps), he or she will be able to define the appropriate task. In addition, if the maintenance history is efficient and reliable, proactive maintenance can be carried out with the statistical calculation of the component's life (AE), using Reliability and Life Calculation tools through appropriate statistical methods, such as analysing the Weibull distribution).

Another important point to take into account is the experience of maintenance operators. As there is a great deal of interaction between the team and the equipment, an analysis with the maintenance team can also provide a lot of information for the analyst who is going to implement the system (SMITH; HINCHLIFFE, 2004).

2.2 RELIABILITY

"Reliability is the probability that a device will perform a specific function satisfactorily for a specified period of time, under certain operating conditions" (SMITH; HINCHLIFFE, 2004, p.1080-1094). According to the terms described in this definition, reliability is a design attribute of the system, equipment or component, and it is up to maintenance to act preventively and in the most appropriate way possible, in the hope of keeping the system as a whole at least at the reliability stipulated in the project.

According to the same authors, in MCC, reliability analysis is only carried out on components that perform a relevant function in the system. At the time of a study, historical data on previous failures is analysed and, if this study indicates a tendency towards failure in relation to a parameter (e.g. time of use), this component can be targeted for preventive maintenance.

"Reliability is the ability of an item to perform a required function under specified conditions during a given time interval" (ABNT NBR 5462/1994), and is defined for engineering analysis as a probability.

In addition, the reliability of one component of a system directly affects the reliability of the system as a whole.

The more complex the system, the lower its reliability. It is therefore necessary for each component to intrinsically have a high degree of reliability and be as simple as possible, so that total reliability is equally high (SMITH; HINCHLIFFE, 2004).

According to the same authors, in practice, the best way to maintain a system with high reliability is to keep it simple and with high reliability for each component separately.

Based on historical data and statistical reliability calculations, we can calculate B10 - fatigue life index with 10 per cent reliability - MTTF - mean time to failure - or MTBF - mean time between failures (PRIDE 2008). These indicators help the MCC implementation analyst to define the tasks necessary for its correct implementation and when to carry them out.

In order to carry out reliability analysis, the most up-to-date technique uses various tools, the one used in this study being AE - life analysis, using the Weibull distribution.

2.3 LIFE TIME ANALYSIS (AE - *Age Exploration)*

The study of lifetime analysis is an important part of the MCC method, as it can directly affect the definition of the type of maintenance to be used, and is the main basis of proactive maintenance.

> This process, known as Life Analysis (LA), was used by the US Submarine Force in the late 1970s to extend the time between periodic overhauls and replace time-based tasks with condition-based tasks. The initial programme was limited to *FBM (Fleet Ballistic Missile)* submarines, but was continually expanded to include all submarines, aircraft carriers and other combat ships and *MSC (Military Sealift Command)* ships. In addition, the Navy invoked the requirements of this type of study for MCC and monitoring as part of new ship developments (NASA, 2000, p. 1-2).

We also have that: "Lifetime analysis (LA) is a key element in establishing an MCC programme. This analysis provides a methodology for varying the main aspects of the maintenance programme in order to optimise the process" (NASA, 2000 p. 3-48).

The most commonly used tool in lifetime analysis to calculate system reliability, based on past failure history, is the Weibull distribution. "The Weibull distribution is commonly used to determine the probability of failure due to fatigue" (NASA, 2000 p. 3-48).

Three stages must be followed, according to NASA (2000):

Review the technical content to ensure that all failure modes are included in the maintenance tasks.

- Adjust the inspection period by analysing the component's life data.
- Grouping tasks in order to improve inspections and reduce downtime for equipment maintenance.

With historical data from previous maintenance, and with this data behaving in a certain way in relation to, for example, time, the analyst specifies the task to be carried out. The task listed may not always be the replacement of the component, even a disassembly for inspection (FF - Preventive Inspection to Locate the Fault) can be specified, as long as it is combined with another breakdown of the equipment, in order to reduce its economic impact, which is also a key point of MCC.

In the light of the above, we realise the great importance of a systematic and efficient data collection system in plant maintenance, without which no kind of lifetime analysis can be carried out.

CHAPTER 3

VEHICLE APPLICATION

Fleets of vehicles - buses, lorries, tractors, off-highway equipment, for example - are known for having their maintenance under control, as poor maintenance will have serious economic, financial and safety consequences.

Despite this, and despite the fact that all servicing is carried out in accordance with the plan specified by the manufacturer for the vehicle, some faults do occur, with the consequences explained above.

Bus and coach fleets are implementing preventive maintenance based on their own analyses, thereby saving 27% on operations. A study of one such company, which implemented preventive maintenance in order to mitigate faults, found that, in addition to the satisfaction of its customers (the fleet's passengers), it achieved a considerable reduction in costs. This company has complete mileage data on its vehicles relating to events that have occurred during the period and uses it to update its checklist, sometimes generating inspections with shorter periods than those specified by the manufacturer, mainly due to the conditions in which its vehicles operate (verbal information).[1]

The MCC system proposed in this study should guide preventive overhauls with regard to the most important parts to check, and will be updated based on information previously compiled on vehicles with similar characteristics in terms of mileage (space travelled), based on faults that have occurred in the past and that will be analysed in the fleet that is in the control domain.

A similar study has already shown that, using appropriate statistical techniques - more specifically, using the MCC technique - it is possible to correlate past data with components that are likely to fail (NASA, 2000).

Vehicles use a large number of mechanical components and are therefore subject to wear and tear per kilometre. These types of components have a failure behaviour (FDP - failure probability density function) according to the traditional "bathtub curve" pattern

[1] News provided by Walter Alves de Oliveira, engineer responsible for the fleet, at the III Congress of the Vehicle Repair Industry of the State of São Paulo - COMETA Preventive Maintenance, on 8 October 2008.

(Graph 1, curve type E) or with a tendency to show an inflection in the curve (Graph 1, curves A and F). This would justify a preventive, proactive maintenance task or a fault search (FF - preventive inspection), scheduled according to a certain mileage.

Other components, such as electrical and electronic components, whose FDP behaviour is dictated by curves similar to those of types B, C and D (Graph 1), are not subject to tasks determined by mileage and must be treated differently during maintenance.

For components subject to preventive maintenance, preventive inspection (FF) or proactive maintenance, the MCC system can be a valid option, as long as historical maintenance data is present.

The CCM will help the analyst, as seen in Chapter 2, to specify the appropriate tasks for each component in order to mitigate the occurrence of faults that could paralyse the vehicle, consequently affecting safety and causing financial losses.

The normal preventive inspections recommended by manufacturers are mainly based on tests and the calculated life of components used in other systems, and are therefore not adapted to the extreme use of fleet vehicles. After these inspections, vehicle failures can still occur, either due to a failure to correctly predict the mileage of component failure, or due to failures inherent in the inspection process, i.e. generated by labour errors on the part of the maintenance operator.

But even if the user is able to understand that the post-preventive inspection failure may not have been caused by the maintenance operator, the vehicle has failed and this causes disappointment, as well as loss of money and time. It's even worse if the failure occurs in a remote location or with little access to repairs and resources, a risk that the vast majority of off-road vehicles face, unlike vehicles used in fleets.

The user's perception of a preventive overhaul is that of a complete and thorough inspection. However, due to time and cost issues, it is not feasible to dismantle and check all the vehicle's systems and components. It is also worth pointing out that the inspection is like an audit of the vehicle at a given moment, and the fault may occur soon afterwards. And this is the fault that the implementation of MCC aims to avoid, at a lower cost than conventional preventive maintenance systems.

CHAPTER 4

APPLICATION EXAMPLE

To begin the application example, the engine of the Land Rover Defender models 90, 110 and 130 was chosen, with a more in-depth focus on its water pump, the main component of its cooling system.

A survey carried out at the company studied showed that the main complaint from users of a fleet with monitored maintenance was overheating of the engine, mainly due to a fault in the water pump, which caused financial losses and reduced their degree of reliability in the vehicle, which is essential in this case, as most users drive it in inhospitable locations and with great difficulty in retrieving it in the event of a breakdown.

4.1 ENGINE HISTORY

The 300 *Turbo Direct Injection* (Tdi) engine was developed from the 200 Tdi engine by Land Rover in England. "200" was, at the time of development, the torque in lbf x ft desired for the engine, which eventually reached 199lbf xft.

The 200 Tdi was the first fast engine - above 4,000 revolutions per minute - to generate power above 100BHP, reaching 111BHP.

With a block derived from older engines (developed in the 1960s) and an externally developed cylinder head, this engine equipped Land Rover vehicles from 1989 (launch of the 200 Tdi) until 1999, when the Td5 engine was launched to meet the demands of European EURO III environmental legislation.

In 1993, Land Rover sold the rights to manufacture the 300 Tdi engine to the Brazilian company MAXION, part of the IOCHPE group, which is now owned by MWM International.

In Brazil, the Defender vehicle in its 90, 110 and 130 versions was built from 1999 to 2005 using the Euro II engine, derived directly from the 300 Tdi engine.

With improvements to the pistons, cylinder head and turbine, the national 300 Tdl engine was able to meet the Brazilian Euro II standard, hence its name.

The engine manufactured by MWM International, which is the subject of this study, was used in Mercedes Benz Sprinter vehicles, Ford's Ranger and Troller, Chevrolet's S10,

Land Rover's Defender, CROSS LANDER - derived from the Aro company, now discontinued - as well as various off-road vehicles and equipment manufactured in Brazil.

4.2 ENGINE AND COOLING SYSTEM CONSIDERATIONS

DEFENDING

The 300 TDi is a water-cooled 4-cylinder turbodiesel engine with a volume of 2.5 litres.

The cast iron cylinder block is integrated into the crankcase and incorporates directly mandrel-bored cylinders.

The engine head is made of cast aluminium alloy and supports a conventional rocker shaft and tappets that operate two valves per cylinder from a single camshaft.

The graphite-coated aluminium alloy pistons are equipped with two compression segments and one lubrication segment, and are attached to the connecting rods by semi-floating piston pins, which are press-fitted into the connecting rod foot bush.

A turbulence chamber is incorporated into the piston crown to facilitate combustion (direct injection).

When this engine is fitted to the Defender vehicle, a *polyvine* belt with a width of seven channels and measuring 1580mm drives the alternator, the water pump, the

power steering pump and fan impeller, and has a spring-loaded mechanical stretcher for automatic stretching.

A second four-channel belt measuring 1430mm drives the air-conditioning compressor, with a manual mechanical tensioner and a positioner pulley.

In diesel cycle engines, the cooling system is responsible for maintaining the engine's temperature within the optimum specification, with the fluid at temperatures between 86° and 105°C.

Temperatures below this level cause consumption and emissions to be higher, at which point a thermostatic valve (Figure 3) kicks in, partially or totally removing the flow of cooling fluid , preventing it from circulating in the radiator (Figure 4) and therefore heating up the engine more quickly.

Figure 3 - Thermostatic valve

Figure 4 - Radiator and crankcase oil cooler

In addition to the thermostatic valve, a viscous actuator (or viscous coupling) acts on the propeller (Figure 5), decreasing or increasing the synchronisation of the propeller's rotation in relation to the pulley. When the viscous coupling detects that the radiator temperature is above that specified, a valve is closed and the recirculation of the coupling's internal fluid is decreased, increasing the rotation of the propeller or, otherwise, the rotation is decreased. The viscous coupling enables considerable fuel savings on high-speed journeys, where the front air is largely responsible for cooling the engine.

Figure 5 - Viscous coupling and propeller

In an engine with a turbine - commonly known as a turbo compressor - the engine's lubricating oil is used to lubricate the central part of the turbine, and this causes a considerable increase in temperature, which can lead to its degradation. It is therefore essential to install an engine oil cooler in order to keep the temperature of the lubricant within acceptable levels.

In the case of the Defender, the crankcase oil cooler is coupled to the coolant cooler (Figure 4), with a portion of the fluid circulation being diverted via a venturi (Figure 6). The venturi (or Y-valve) is also partly responsible for removing air from the top of the engine (cylinder head) in order to avoid hot spots during firing.

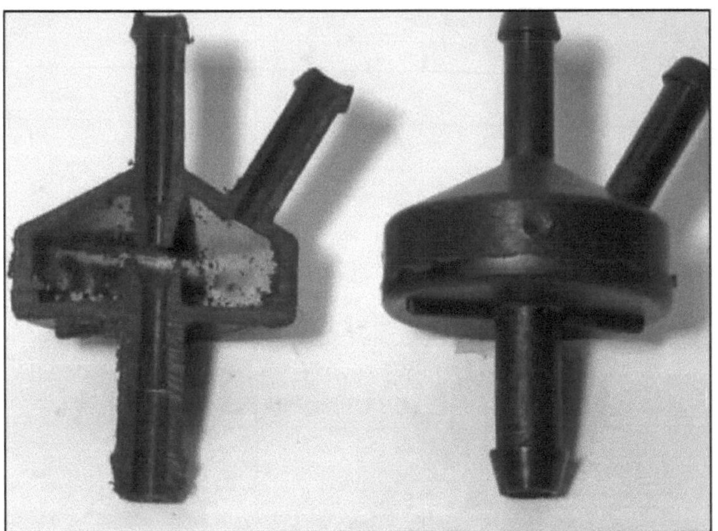

Figure 6 - Venturi: internal view (left, cut) and external view, original part

In order to improve burning, another device, commonly called an *intercooler* (Figure 7) cools the air before it enters the intake manifold, also helping to cool the engine.

Figure 7 - Intercooler

The air coming out of the turbine, compressed to around 1 kgf/cm^2 and with a temperature of around 250°C, is cooled by a radiator next to the fluid radiator, rejecting a large part of the heat that was acquired in the compression process, and the air is admitted to the manifold with a temperature of around 80°C.

Another device that forms part of the cooling system is the expansion vessel, responsible for the system's internal pressure, which is maintained at 1 kgf/cm^2 in order to keep the boiling temperature within an acceptable limit even at high altitudes, preventing the fluid from boiling at lower temperatures, generating engine heating problems.The expansion vessel lid is responsible for keeping the pressure within this level and also for the vent, allowing air to enter after the system has cooled and its consequent decompression.

4.3 DEFENDER WATER PUMP

The water pump, the main object of this study, has the function of supplying fluid flow at a rate of 178 litres/hour at 3800 rpm and had several faults in its impeller.

Once failures have occurred, it is difficult to find out whether the pump failed because the motor overheated or vice versa. In order to determine this, several pumps were dismantled before heating up and it was found that the impeller had cracked (Figure 8) before the motor overheated. Once cracked, the rotor separates from the shaft (Figure 9), ceasing to perform its main function of pumping fluid from the engine to the radiator in order to cool it. The most important was the impeller failure (Figures 8 and 9), which is impossible to detect without a preventive inspection (FF) and complete disassembly of the pump.

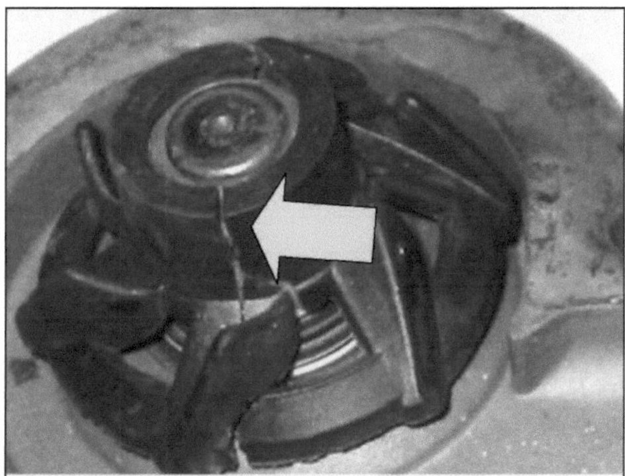

Figure 8 - Detail of the impeller of a damaged pump (internal view)

Figure 9 - Detail of a destroyed rotor, separated from the shaft

The second most important type of failure was fluid leaking through the retainer, shown in the external view (Figure 10) and in the internal view in the detail of the shaft with the damaged retainer (Figure 11). The third failure mode was a bearing problem, but before the bearing breaks, it makes a characteristic noise, which alerts the user to the failure.

Figure 10 - Detail of fluid leaking through the retainer hole (external view)

Figure 11 - Detail of shaft with damaged retainer (internal view)

Both the second and third failure modes were reported by the users themselves, which indicates that they are easy to detect, although they also require changing the water pump, as it is only sold complete.

There are doubts as to whether the two less important failure modes have any correlation. A seal failure causes fluid contamination in the bearing housing, damaging it (by removing its lubrication), which can be the cause of bearing failure.

The time that elapsed between the failure of the seal and the failure of the bearing cannot be determined; the fact that correlates the two failures is that there was no record of the bearings failing without the seal failing.

4.5 CASE STUDY EXAMPLE

The example case study was carried out by implementing the seven stages of the rapid MCC method and defining the focus component of this study in order to delimit the work and guide future actions. Due to the criticality of the defect and the

The water pump was the component chosen for the case study. Failure of this component seriously damages the engine, requiring expensive repair work for the owner, estimated at around R$6,000.

Subsequently, a detailed survey was carried out of the components that failed over the years 2004 to 2008 in the vehicles repaired, related to their mileage. This survey covered the components replaced on a total of 2164 journeys in the aforementioned period (registered Service Orders), data which was entered into a database and compiled, as shown in the example below (Figure 12).

Figure 12- Survey of work orders (WO) - (2004-2008) Entering data in a summary spreadsheet for later compilation

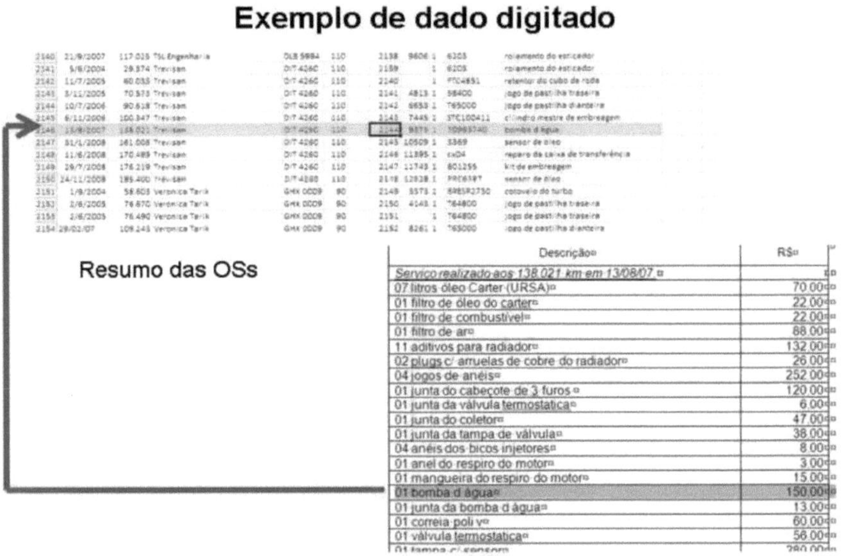

Exemplo de dado digitado

2140	21/9/2007	117.015	TSL Engenharia	DLB 5984	110	2138	9606	1	6203	rolamento do esticador
2141	5/6/2004	29.574	Trevisan	DIT 4260	110	2139		1	6203	rolamento do esticador
2142	11/7/2005	60.035	Trevisan	DIT 4260	110	2140		1	FTC4851	retentor do cubo de roda
2143	3/11/2005	70.573	Trevisan	DIT 4260	110	2141	4813	1	58400	jogo de pastilha traseira
2144	10/7/2006	90.618	Trevisan	DIT 4260	110	2142	6653	1	T65000	jogo de pastilha dianteira
2145	6/11/2006	100.347	Trevisan	DIT 4260	110	2143	7445	1	STC100411	cilindro mestre de embreagem
2146	13/8/2007	138.021	Trevisan	DIT 4260	110	2144	9379	1	[illegible]	bomba d água
2147	31/1/2008	161.008	Trevisan	DIT 4260	110	2145	10509	1	3369	sensor de óleo
2148	11/6/2008	170.489	Trevisan	DIT 4260	110	2146	11395	1	cxD4	reparo da caixa de transferência
2149	29/7/2008	176.219	Trevisan	DIT 4260	110	2147	11743	1	801255	kit de embreagem
2150	24/11/2008	185.400	Trevisan	DIT 4260	110	2148	12828	1	PRC6397	sensor de óleo
2151	1/9/2004	58.603	Veronica Tarik	GHX 0009	90	2149	3573	1	SRESR2730	cotovelo do turbo
2153	2/6/2005	76.870	Veronica Tarik	GHX 0009	90	2150	4148	1	764800	jogo de pastilha traseira
2155	2/6/2005	76.490	Veronica Tarik	GHX 0009	90	2151		1	764800	jogo de pastilha traseira
2154	29/02/07	109.243	Veronica Tarik	GHX 0009	90	2152	8261	1	765000	jogo de pastilha dianteira

Resumo das OSs

Descrição	R$
Serviço realizado aos 138.021 km em 13/08/07	
07 litros óleo Carter (URSA)	70.00
01 filtro de óleo do carter	22.00
01 filtro de combustível	22.00
01 filtro de ar	88.00
11 aditivos para radiador	132.00
02 plugs c/ arruelas de cobre do radiador	26.00
04 jogos de anéis	252.00
01 junta do cabeçote de 3 furos	120.00
01 junta da válvula termostatica	6.00
01 junta do coletor	47.00
01 junta da tampa de válvula	38.00
04 anéis dos bicos injetores	8.00
01 anel do respiro do motor	3.00
01 mangueira do respiro do motor	15.00
01 bomba d água	150.00
01 junta da bomba d água	13.00
01 correia poli v	60.00
01 válvula termostatica	56.00
01 tampa c/ sensor	280.00

Service Order (SO)

The forecast for the implementation of preventive inspection of this component was carried out using an appropriate statistical method. It was decided to carry out a preventive inspection every 20000km on the example case study component from the end of 2008. Of the vehicles inspected, most had a defect in the component, proving that, despite the higher cost of preventive inspection, there were savings for the owner and a reduction in the frequency of heating problems.

This study proposed two solutions to the problem found in the component under study: the first consisted of sending a study to the supplier, informing them of the problem and requesting the necessary corrections, and the second (section 4.8) referred to the assembly of a specially developed device with the purpose of informing the user of the level of fluid present in the cooling system on the vehicle's dashboard, which could prevent overheating in the event of a fault occurring.

The information needed to analyse the data in this case study is available for consultation at .

4.6 THE IMPLEMENTATION OF MCC IN THE CASE STUDY EXAMPLE

As seen above, the following seven steps must be taken when implementing the MCC:

1) *Establish the scope of the study,* to clearly define the limits of the work.

The engine water pump (Figure 13) was chosen as the object of study because it has a critical fault that is difficult to detect and expensive to repair.

Figure 13 - Defender's new water pump (internal view)

This fault was classified as critical, mainly because the difficulty in detecting its occurrence could lead to damage to the cylinder head (Figure 14). In addition, if the user does not realise the fault in time to avoid it, the engine will have to be completely rectified, generating a high cost.

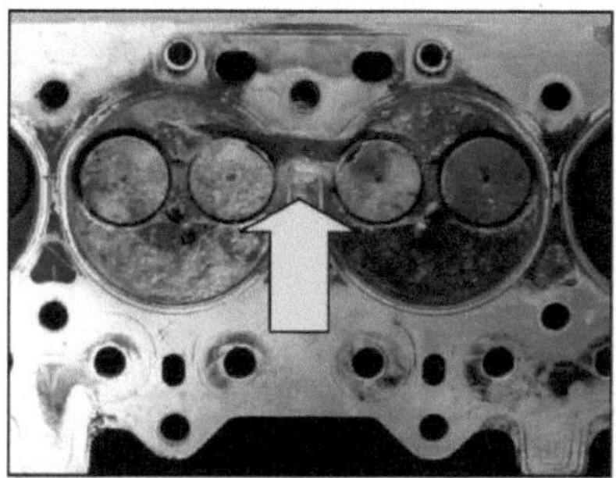

Figure 14 - Head damaged beyond repair due to excess temperature (internal view)

In March 2010, the cost of rectification was estimated at between R$2500, exclusively for cylinder head rectification, and R$15000 for a complete engine rectification, with the inspection and replacement of the pump costing between R$200 (inspection only) and R$600 (inspection plus replacement).

2) *Identify the interfaces in* order to clearly establish which inputs and connections will not be studied.

A detailed flowchart of information about the Defender's engine cooling system was drawn up (Figure 15). It was considered that although the water pump is made up of a housing, shaft and rotor, it is sold as a single assembly, i.e. it is not sold in parts, and is therefore considered to be an isolated component (Figure 13).

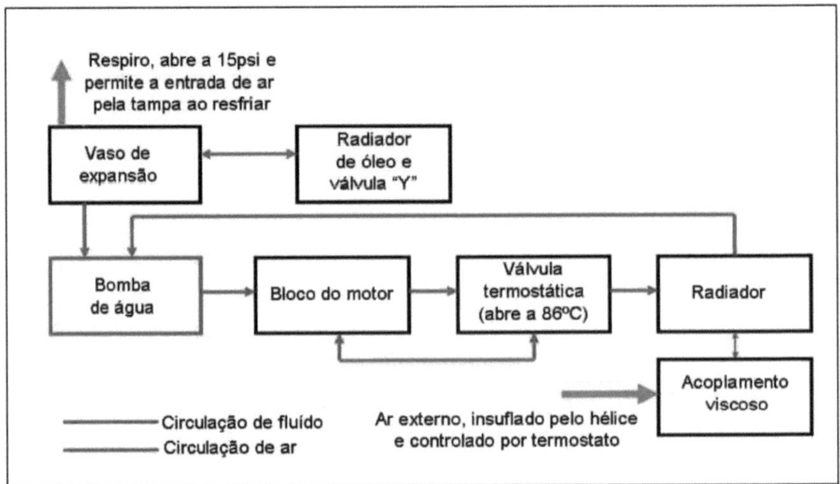

Figure 15 - Defender cooling system flowchart

3) *Specifying important functions.* MCC seeks to preserve only the most important functions of a system or piece of equipment. The main function of the water pump is to provide the fluid flow needed to cool the engine (178 litres/hour at 3800rpm), without which the engine will overheat and cause serious damage to the vehicle's other components.

4) *Identify the dominant failure modes.* For evaluation purposes, only the main functions/systems will be studied.

The dominant failure mode identified by the FMEA carried out is rotor rupture (Figures 8 and 9), which prevents the flow of coolant. Other failure modes, although they do exist, are easier to detect and can be diagnosed by the user without the intervention of the workshop, which is left to repair. In all cases, the only option is to replace the pump, as it is not sold for repair.

5) *Identify critical failure modes.* The consequences of failures are assessed for each dominant failure in order to determine its severity. Non-critical failure modes are not studied by MCC.

Rotor failure was listed as the most critical due to its serious consequences, as well as the fact that it is impossible to detect without dismantling the rotor.

The FMEA analysis is important for defining the main failure mode to be analysed. When drawing up the FMEA for the water pump, the following procedures should be carried out:

- *Determine the product's functions and characteristics*

The main function of the water pump is to provide a flow rate of 178 litres/hour at 3800 rpm. Its main characteristics and the study of its fault types were determined in section 4.3 of this study.

- *Determine the type of potential failure for each function.*

As the function of the pump is only one - to provide flow, but there are three different reasons why this flow is not provided, three failure modes were listed: impeller breakage, fluid leakage and bearing breakage.

- *Determine the effect of the type of fault.*

In the three faults described above, the potential effect of the faults is the same: engine failure. The difference between the faults is that the first one is hidden from the user's eyes, while the last two can be noticed by the user.

- *Determine the possible cause of the fault.*

The first failure is due to problems with the rotor material, the second is caused by a faulty mechanical seal, and the third is due to a faulty or broken bearing

- *Determine the current controls to mitigate the fault.*

In the first case, as it is possible to carry out the survey and calculate the estimated life of the pump (AE), a periodic inspection (FF) was proposed in order to check the condition of the impeller; if it showed signs of cracking, the pump should be replaced. In the second and third cases, if the symptoms appeared (leak detected or noise detected, respectively), the pump should be replaced.

Assess the risks of the fault and classify the Severity (S), Occurrence (O) and Detection (D) indices according to the tables.

Tables 1, 2 and 3 were used in all the cases in this study. As the risks to the engine are the same for all three faults, the Severity index was classified as the same (S = 8) for each. With regard to Occurrence, the indices were calculated from the table based on the data contained in the company's OSs: (0 = 9 for the first failure mode and 0 = 4 for the others). With regard to Detection, the first failure mode cannot be detected without dismantling the pump, and D = 10 was assumed. For the second and third failure modes, D = 5 was assumed, as the user is able to recognise the failure themselves (Figure 10). With regard to Severity, it is worth remembering that if the user is travelling in dangerous places, it could be assumed that there is a risk to their safety, and a factor of S = 9 could be assumed.

- *Calculate the risk index by multiplying the three indices determined above with each other,* as shown in Table 4 - FMEA Form.
- *Propose improvement measures.*

For the failure mode listed as the main one in this study, the pump life (AE) was calculated. Then, estimating BX=0.1% (life index with a 0.1% probability of failure), the result was 27498km (Table 5). As the vehicle is inspected every 10000km, the pump (FF) was opened and inspected every 20000km in order to mitigate the other two failure modes. The pressurisation of the cooling circuit was listed at the 40000km periodic inspection and the inspection of the aggregate bearings (all) was scheduled every 10000km.

- *Prepare and display this data on an appropriate form.*

In order to organise the information properly, presenting this information on a single form offers advantages in terms of analysis, summarising all the information collected in a single document (Table 4).

Failure Mode and Effects Analysis (FMEA) Manutenção

Produto: Bomba dágua 70993740 conjunto completo
Aplicação: Land Rover Defender 90/110/130
Preparado por: Luiz A.C.F.Mo
Data revisão FMEA: 23/abr/09

Item/ Função	Modo de falha potencial	Efeito potencial da falha	Sev.	Causa(s) potencial/ mecanismo da falha	Ocorr.	Controles correntes	Detec.	RPN	Ações recomendadas	Resp. e data obj	Resultado das ações: Sev	Occ	Det	RPN
Rotor/ prover fluxo	quebrar	fundir o motor	8	material inadequado	9	nenhum	10	720	abrir e checar o rotor conforme especificação da AE	Specialist	8	1	1	8
Retentor / prover vedação	vazar	fundir o motor	8	selo mecanico defeituoso ou quebrado	4	vazamento detectado	5	160	trocar a bomba	Specialist	8	1	1	8
Rolamento / permitir rotação	quebrar	fundir o motor	8	rolamento defeituoso ou quebrado	4	ruído detectado	5	160	trocar a bomba	Specialist	8	1	1	8

Table 4- Water pump FMEA form illustrating the three failure modes

6) *Identify the dominant causes of failure modes.* Only the dominant causes of critical failures are identified. Only causes of failures subject to preventive maintenance will be studied.

Once the critical failure had been analysed, a survey of historical data (AE) was carried out, and it was concluded that it was possible to implement preventive inspection (FF) for this specific case, since the device fails in a specific time, translated into mileage, which is feasible to calculate, according to the statistical analysis carried out in section 4.7 of this study

Also necessary at this stage is the LTA, illustrated above in Figure 2, which is carried out for each failure mode listed in the FMEA:

1) *Rotor breakage.*

As far as detection is concerned, the failure mode was classified as hidden D; as far as safety is concerned, it does not cause a safety problem; and, as far as interruption is concerned, it causes interruption at a high cost, being classified as B.

Conclusion: "D/B" classification

2) *Leakage through the retainer*

In terms of detection, the failure mode was classified as detectable (it can be detected by the user); in terms of safety, it does not cause a safety problem; and in terms of interruption, it causes interruption at a high cost, being classified as B.

Conclusion: "B" classification

3) *Bearing failure*

In terms of detection, the failure mode was classified as detectable (it can be detected by the user); in terms of safety, it does not cause a problem of this type; and in terms of interruption, it causes interruption at a high cost, being classified as B.

Conclusion: "B" classification

According to the guidelines in section 2.1 of this study, the three failure modes lead to preventive maintenance for failure inspection (FF), which is an applicable and effective task for mitigating failures.

In order to determine the useful life of the pump in relation to the "rotor breakage" failure mode, a statistical study was carried out, based on the analysis carried out and explained in section 4.7 of this study.

The term BX is a term used by bearing manufacturers in the early days of reliability engineering to refer to the time in which X percentage of units in a population will have failed. More specifically, "B10" is an index that refers to the time in which 10 per cent of bearings can fail (NASA, 2000). BX is an index that uses the probability of failure calculation instead of reliability, the latter of which is normally used for warranty time calculations.

To determine the mileage at which a preventive inspection should be carried out, a life analysis (LA) is carried out, which leads to the conclusions illustrated in tables 5 and 6 below:

Table 5 - Calculation of the space travelled for BX for 0.1%

BX = 0.1%	**Km**
Superior =	30529
Time =	**27498**
Lower =	25074

As can be seen in Table 5, in order to obtain a BX index of 0.1%, the average space travelled would have to be 27498km.

Table 6 - Calculating the average life of the Defender water pump

Average life	**Km**
Top (km)=	418930
Average life (km)=	**383670**
Lower (km)=	351480

As can be seen in Table 6, the average life of the pump was estimated at 383670km

.

It's worth noting that data acquisition is continuing at the company, with the aim of further increasing the accuracy of the pump life calculation.

7) *Select maintenance tasks.* Using logical decisions, maintenance tasks are listed in order to prioritise the dominant and critical causes of failure. Design changes can be considered at this stage, allowing the equipment / system to operate without failure.

The Defender manufacturer stipulates that preventive inspections of the vehicle under study take place every 10000km. Due to the results obtained in Table 5 (27498km), it was specified that the pump inspection for the "rotor breakage" failure mode (FF) be carried out every 20000km. As for preventive inspections for the "rotor leakage" (FF) failure mode, it was specified to be inspected every 40000km with pressurisation of the cooling circuit; for the "bearing breakage" (FF) failure mode, it was specified to be inspected every 10000km.

4.7 STATISTICAL ANALYSIS OF THE CASE STUDY EXAMPLE

To determine the estimated life analysis (LA) of the pump, a survey was carried out on 2162 work order forms from January 2004 to December 2008, which was then submitted to an appropriate statistical system. From this survey, 127 pumps showed rotor failures, representing around 6 per cent of all tickets.

The 2162 passes represented a total of 447 vehicles that had their water pumps inspected during this period.

Taking into account that 9,377 Defender vehicles were sold from 1992 to 2008 (Table 7), that 8,737 of these vehicles use the 300 Tdi engine (manufactured between 1995 and 2006) and that, of these, those manufactured between 2002 and 2006 even use the pump that is the subject of this study, there are a total of 4,110 vehicles on the Brazilian market subject to the problem that is the subject of this case study example.

Table 7 - Number of vehicles sold (1992-2008)

Mote lo	1992	1993	1994	1995	1996	1997	1998	1999	2000	2001	2002	2003	2004	2005	2006	2007	2000	Accumulated
Oetenoer	11	65	170	36	301	521	706	695	1.211	1.157	1.677	857	796	639	141	27	367	9.377
Discovery	12	39	73	9	102	244	282	182	115	196	171	161	225	393	788	1.162	1.397	5.548
Freelander	0	0	0	0	0	0	0	0	0	0	0	117	672	1.002	747	1.717	2.267	6.522
R.Rover	4	6	10	0	1	55	40	53	17	16	9	69	78	62	33	54	65	572
R.R. sport														78	246	356	477	1.157
rotai	2/|	110|	253|	45|	4U4|	820|	1.0281	930|	1.9431	1.3601	1.8571	1.204|	1.7711	2.1711	1.9 5S>|	3.3141	4.573	23.176
oetenoer acum.	11	76	246	282	583	1.104	1.810	2.505	3.716	4.873	6.550	7.407	8.203	8.842	8.983	9.010	9.377	
Defender bras.								695	1.906	3.063	4.740	5.597	6.393	7.032	7.173			
Bomoa rotor der											1.677	2.534	3.330	3.969	4.110			

Source: Land Rover do Brasil, 2009[2]

According to the author's estimate, around 40% of these vehicles were sold in the state of São Paulo (totalling 1,644 vehicles), 60% of which were in the capital, making up 988 vehicles.

This data, according to the estimate, indicates that the workshop serviced 45 per cent (447 vehicles) of the manufactured fleet circulating in the capital of São Paulo between 2004 and 2008, the period of the statistical analysis.

Below are the graphs obtained by calculating them using a computerised system (RELIASOFT), which provided the basis for the statistical analysis.

Graph 2 shows the distribution of the probability of failure in the space travelled by

---

[2] Information provided in a personal interview by the commercial manager of Land Rover do Brasil in 2009.

the vehicles that took part in the application example. It was concluded that the estimate calculated by the system is consistent with the practical data collected.

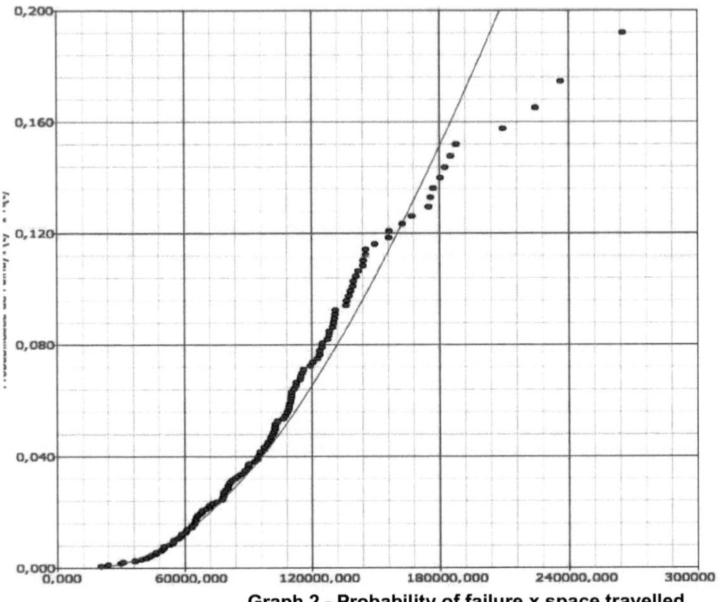

Graph 2 - Probability of failure x space travelled

Graph 3 shows the similarity of the behaviour of the failure probability density function by space travelled with the F-type curve in Graph 1, suggesting preventive inspection or proactive maintenance with life analysis (LA).

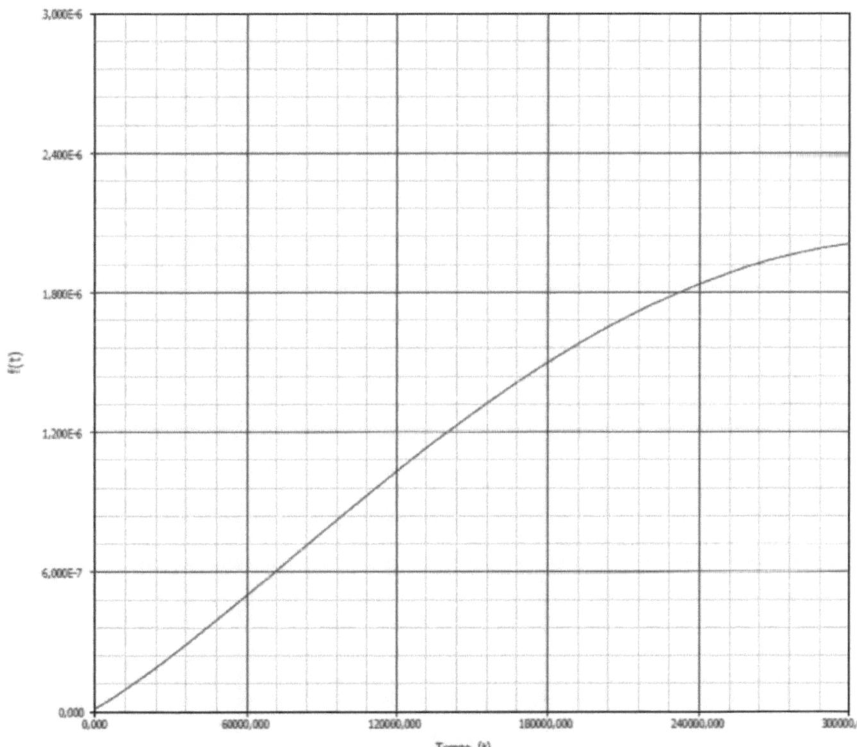

Graph 3 - Fault probability density function x space travelled

Graph 4 shows the failure rate in relation to the space travelled by the vehicles in the example application. It can be seen that this rate increases in relation to the space travelled.

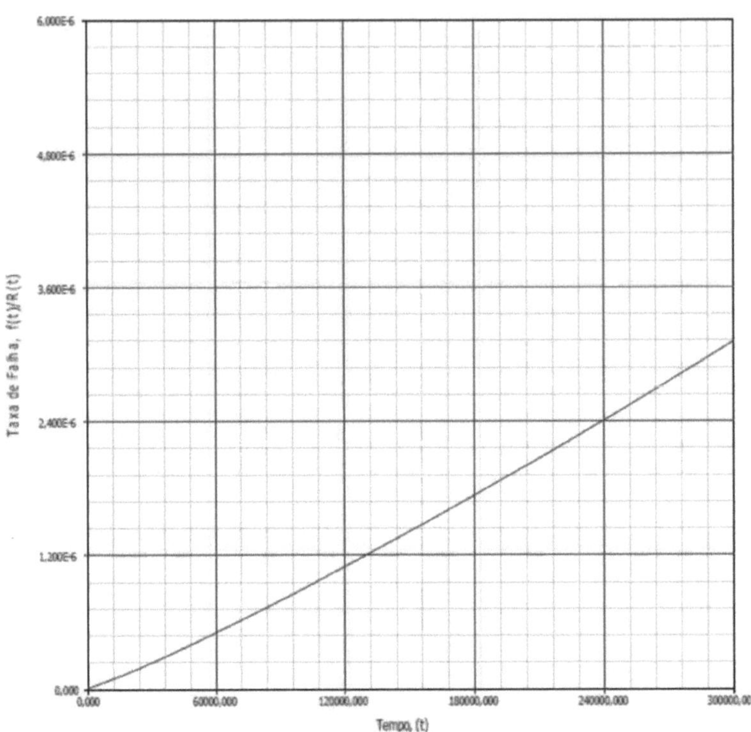

Graph 4 - Failure rate x space travelled

Graph 5 shows the decrease in the reliability of the space travelled in the vehicles analysed. It was also concluded that the estimate calculated by the system is consistent with the practical data collected.

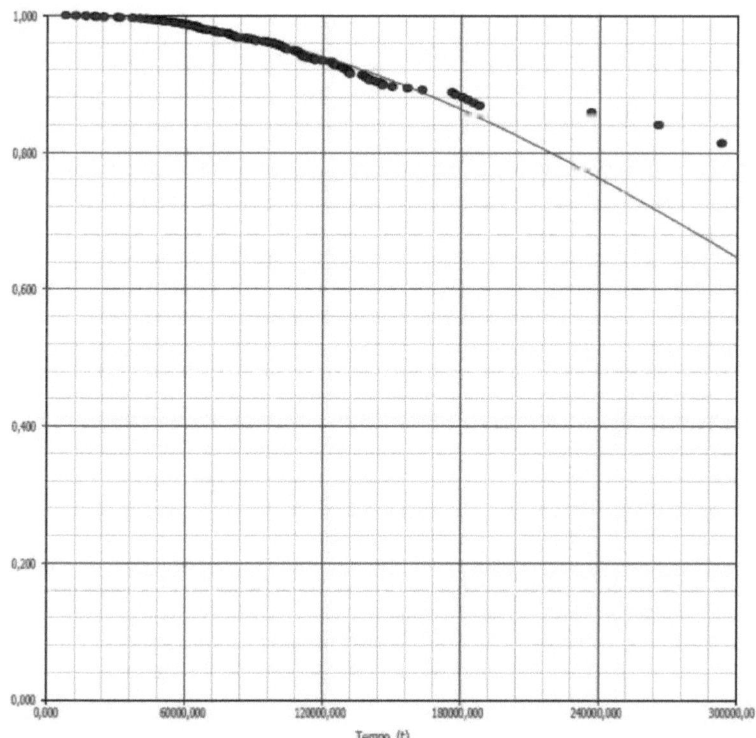

Graph 5 - Reliability x space travelled

Graph 6 illustrates the space travelled by vehicles that failed over the test period compared to those that didn't (suspensions).

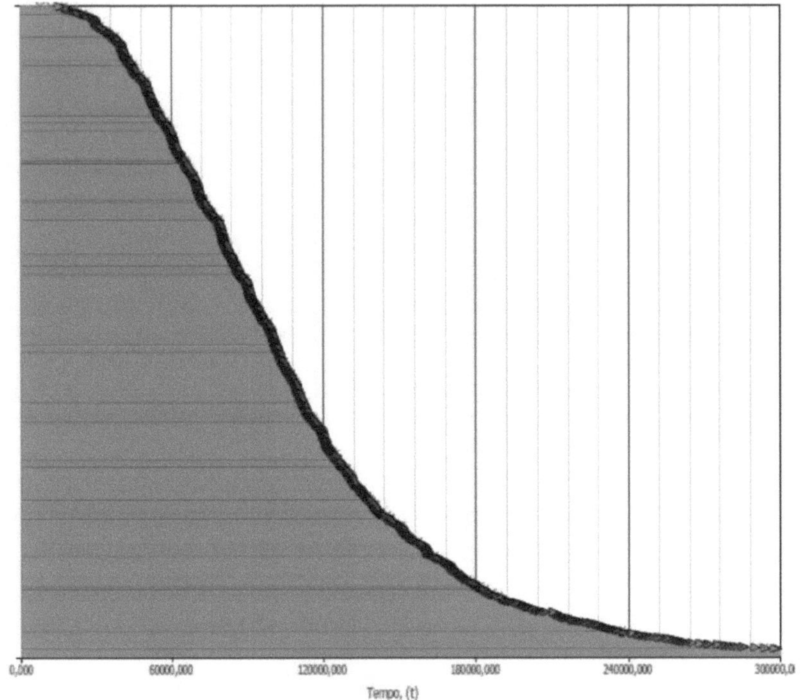

Graph 6 - Line faults and suspensions x space travelled

4.8 STEPS TAKEN IN THE APPLICATION EXAMPLE

As a first step to prevent water pump problems, this study proposed that Defender users install a water level sensor (Figure 16) connected to an electronic system.

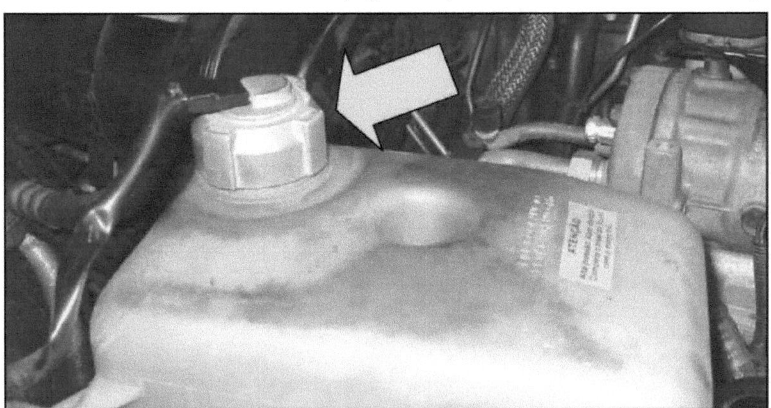

Figure 16 - Expansion tank with lid and level sensor

Because the failure of the water pump impeller generates an increase in temperature and pressure in the fluid, it boils and is expelled through the cap, which depressurises the system when the pressure reaches 1kgf/cm^2) causing, in the worst case scenario, the 12 litres of fluid in the system to be expelled, forcing the engine to work without fluid, causing certain failure in the cylinder head (Figure 14) or reaching the lower part of the engine (pistons and cylinders) with even greater damage.

The sensor is triggered when just one litre of fluid is missing, emitting a visual and audible signal on the vehicle's dashboard, alerting the operator to the situation, which is monitored in real time.

The second step suggested was to contact the supplier of the water pump, asking for changes to its design. This study verified the need to develop a new pump with a metal impeller - as previously supplied (Figure 17) - or to improve the material or production process of the impeller currently supplied.

At the time of this study, the pump supplier had not commented on the suggested measures.

The pump shown in Figure 17 has only two failure modes, seal leakage and a bearing problem, making detection possible for the user.

Figure 17 - Water pump, pre-2002 model with metal impeller (internal view)

A national attempt to improve the impeller by manufacturing a metal impeller using another process - such as casting - has proved problematic for the time being. Figure 18 illustrates a pump with a damaged cast metal impeller, manufactured by a local supplier. The use of the cast metal impeller has not proved efficient, as it breaks in a similar way to the nylon impeller and the water pump assembly on the parallel market costs more than the original.

Figure 18 - National water pump with damaged cast metal impeller (internal view)

4.9 MOTIVATIONS AND CRITICISMS OF THE EXAMPLE CASE STUDY

One of the motivations for choosing the system studied arose from the need for users operating the vehicle under study to increase reliability in their vehicles, since faults can mean a safety problem when they occur in remote locations or further away from the city, which are known to be the favourite places for travelling in off-road territories.

Another important motivation was to reduce the cost of maintenance for users, allowing them to anticipate the failure, reducing possible future expenses and improving confidence in the vehicle after repair, as mentioned above.

With regard to criticisms of the MCC system, in the case of the specific study object there is no possibility of a test before dismantling. The user must therefore bear the cost of disassembly for verification, and if the device (water pump) has not yet ruptured its impeller, labour and some components (coolant and gasket) will have to be charged for, causing an outlay for the owner.

Another criticism of the MCC system is that it is impossible to foresee the maintenance of devices that depend on a high level of influence from the driver. For example, the clutch system - specifically, the clutch disc, pressure plate and bearing assembly - which depends on the way the vehicle operator drives it, cannot be maintained with the MCC system, since space travelled and wear are not related, therefore reducing the chance of success in predicting its failure. On the other hand, for fleet management, this data could indicate that the way the user is driving the vehicle is not the most correct, requiring specific training.

In the statistical study of the example case study, some problems with data acquisition may have occurred, such as an erroneous note on the service order form of the vehicle's mileage, or the fact that the user changed the component at another workshop without proper communication, which would alter the data and consequently the component life calculations.

Failures in data acquisition due to inaccurate annotation of the failure mode may also have occurred, leading to the calculation of the useful life of the pump being decreased, an error that would actually be decreasing the useful life of the pump - a fact that would benefit user safety, despite the increase in cost.

The information that the water pump was changed from a metal impeller to a nylon one in 2002 creates an inaccuracy in the statistical survey, which indicates that approximately 45 per cent of the vehicles that circulate in the municipality of São Paulo have passed through the company that is the subject of this study. However, this data was not taken into account when calculating the useful life of the pump.

Implemented at the beginning of 2009 in the company, the system was able to remove from more than 100 vehicles pumps with cracked impellers (Figure 19), which were in danger of causing serious damage to the engines, saving the owners approximately R$ 24,000,000, taking the average unit repair as R$ 6,000 per rectification.

Figure 19 - Water pumps with damaged impellers

A study similar to this one would have to be carried out on other vehicle fleets to see if there is a space travelled where the probability of failure increases, as proven in this study.

Another important observation is that the vehicles to be maintained with this type of maintenance must have strictly monitored records, because if for any reason these are lost or inaccurate, the system's implementation will certainly fail.

CHAPTER 5

CONCLUSIONS

In vehicle fleets that have their maintenance monitored, the implementation of the MCC system comes at virtually no cost, as long as the data to be used in the life analysis is available, and few resources are needed to enter it into an aggregate analysis system or migrate it directly to the statistical calculation system, such as the one used in this study.

In order to also benefit from the experience of the MCC system analysts, which is essential for the success of the implementation, it is suggested to use quality techniques (CCQ - Quality Control Circle, for example) to discuss the problems that are occurring most frequently in the fleet and then analyse them, prioritising them on the agenda of other meetings. For example, if at a meeting one of the mechanics mentions a wheel hub bearing problem, a past survey of this item could be carried out in order to determine whether this failure is a one-off or whether it will be a problem faced by the entire fleet in the near future.

The proximity between the MCC analyst and the maintenance operators is also important so as not to jeopardise the analysis and recording of failure modes. For example, the water pump on the vehicle in this study has three failure modes; a suitable form should be drawn up and the operators instructed to mention on the form that will be used for the record precisely which failure mode was present on the part to be replaced or, optionally, to pass all the parts on for a final inspection by the MCC analyst, which, however, is usually impossible in practice.

Another point worth noting is the so-called generated defect (or fault). This type of fault is usually caused by improper intervention on the vehicle. For example, in the case of this study, if the operator does not top up the coolant level and does not prime the system properly, overheating will certainly occur, not because of a problem with the pump, but because of an error on the part of the maintenance operator.

There is also an interaction between vehicle users and maintenance operators when implementing the MCC programme. It is therefore necessary to adapt the vehicle entry form in the workshop, which contains observations provided by users, transforming it, so to speak, into an anamnesis, or complete description of the symptoms of the vehicle experienced by users. Drawing on all the experience of the MCC programme analyst will help the operators when discovering faults, especially when it comes to the new electronic systems in vehicles that show intermittent characteristics very frequently.

Other important data for detecting faults and resolving them is to check the entire fleet for information on tyre wear and the specific consumption of each vehicle, together with a visual inspection of apparent fluid leaks, the fluid level before leaving and any points that the analyst deems appropriate to include, preferably by having the user fill in an appropriate form, which should also be available to the analyst.

It is recommended for future work to extend this survey to include other vehicle devices and components, and also to other vehicles that may have their useful life estimated, in fleets with controlled maintenance.

In addition, a study using the experience of maintenance mechanics would also be enlightening, since the majority of vehicles on the market are poorly maintained.

A vehicle preventive maintenance programme has been proposed by the trade body SINDIREPA - Union of Repair Companies, to be adopted in the near future with government support; if this happens, its implementation would be facilitated using the technique that is the subject of this study.

A database including all SINDIREPA member companies with information on mileage and component failures of the various vehicles, if compiled and analysed using a specific computer system, could provide both vehicle owners and workshop owners with a powerful tool for preventive maintenance.

In the near future, the government is planning to introduce technical vehicle inspections. If an MCC-type system is seriously implemented, the inspection work would be greatly facilitated and even eliminated in vehicles with an up-to-date preventive maintenance history, a procedure that could even contribute to increasing vehicle safety.

The experience gained from the application example allows us to conclude that CCM can be applied to vehicle maintenance, bringing great benefits to vehicle owners, as long as the maintenance analysis is conducted correctly and the actions resulting from this analysis are consistent with appropriate statistical and maintenance techniques.

REFERENCES

ABERNETHY, Robert B; **The New Weibull Handbook.** 5th edition, Author and Publisher, 2004, passim.

ABNT. NBR5462 - Reliability and maintainability. Brazilian Association of Technical

Standards. Rio de Janeiro, RJ, 1994.

ANNIS, Charles. Weibull Topics. *Weibull Analysis of component reliability.* Available at: http://www.statisticalengineering.com/Weibull/weibull.html. Accessed on: 12 April 2009, passim.

APPS, John, DREW Mick. **Mobile fleet budget & availability forecasting using reliability centred maintenance and reliability block diagrams to forecast fleet availability for mobile equipment with different ages.** Available at:< www.globalreliability.com>. Accessed on: 24 Apr.2010, passim.

CAPALDO, Daniel; GUERRERO, Vander; ROZENFELD,Henrique; **FMEA Failure mode and effect analysis.** 2005-2006. Available at:< http://www.ogerente.com.br/qual/dt/qualidade-dt-FMEA.htm>. Accessed on: 30/04/2009.

DUNN, Sandy. **Reinventing the Maintenance Process:** Towards Zero Downtime. Available at:< http://www.maintenanceresources.com/References/Maintenance> . Accessed on: 26/04/08, passim.

CLARKE, Phil, STEPHEN, Young. **Reliability Centred Maintenance** and HAZOP: Is there a room for both? The Asset Partnership. Pty Ltd 2006. Available at: http://www.scribd.com/doc/7289840/RCMvsHazop. Accessed on: 28 Apr. 2010, passim.

CONGRESSO BRASILEIRO DE MANUTENÇÃO, XXIII, 1-5 September 2008, Abraman Mendes Convention Center. Santos - SP: Brazilian Maintenance Association

CONNOR, Patrick. **Practical Reliability Engineering.** John Wiley and Sons, 4th Edition, 2002, passim.

DEPARTMENT OF MOTORWAYS (DER). **Road statistics.** Available at:< www.der.sp.gov.br/malha/estat mesh/malhaRod07.htm>. Accessed on 26 April 2008.

FLEMING, P. V. & FRANÇA, S. R. R. O. Considerations on the Joint Implementation of MCC and TPM in the Process Industry. In: Proceedings **of the XII Brazilian Maintenance Congress.** Brazilian Maintenance Association (ABRAMAN), São Paulo, 1997, passim.

FMEA Failure Mode Effects Analysis. Available at:< http://www.fmeainfocentre.com>. Accessed on: 30 Apr. 2010, passim.

FMEA Facilitator. Software for preparing FMEA. Available at:< www.fmea.com > Accessed on: 01 May 2010, passim.

FMEA Failure Mode Effects Analysis. Available at: <http://www.daelt.ct.utfpr.edu.br/professores/marcelor/Cap.fmea.pdf>. Accessed on: 01 May 2010, passim.

GERAGHERTY, T. **Achieving Maintenance Cost Effectiveness through the Integration of RCM and TPM Condition Monitoring Techniques.** Translation of: SIQUEIRA, K.T. Available at:< www.confiabilidade.com.br>. Accessed on: 2 May 2010, passim.

LAND ROVER Microcat electronic parts catalogue.2004 version. 1 CD-Rom. Infomedia Ltd, passim.

LATINO, Keneth C. **FMEA A Modified Approach.** Petroleum Refineries Association (NPRA) Maintenance Conference, May 1996, passim.

MATA FILHO, J.N. et al. Reliability-Based Maintenance and Maintenance Cost Control: a successful team in the aeronautical industry. In: **CD-Rom Proceedings of the XII Brazilian Maintenance Congress.** Salvador, BA, 1988.

MATHESON, Thomas D. The air transportation industry: birthplace of Reliability- Centred Maintenance. **Proceedings from the Substation Reliability Centred Maintenance Conference,** Electric Power Research Institute, 1995.

McDERMOTT, Robin E et al. **The Basics of FMEA,** 2nd edition, Kindle edition, 2009.

MOUBRAY, John; **Reliability-Centred Maintenance (RCM II),** second edition, New York: Industrial Press Inc., 1997.

MOUBRAY, John; **Reliability Centred Maintenance,** transl. Kleber Siqueira, S. Paulo:

Aladon. 2000.

National Aeronautics and Space Administration (NASA). **Reliability Centred Maintenance Guide for Facilities and Collateral Equipment,** 2000.

NOWLAN ET AL. **RCM Maintenance Steering Group of the FAA** (1960 as MSG1; 1978 as MSG2; revised as MSG3 in 1993).

PINTO, Renzo Guedes, LIMA, Roberto Carlos Camello. Reflections on RCM integration in a TPM environment. In: **Proceedings of the XIII SIMPEP,** Bauru, SP. 2006, passim.
PRIDE, Alan. **Reliability Centred Maintenance** Smithsonian Institution, 2008.

RELIABILITY Centred Maintenance Project Manager's Guide a.k.a. - The RCM Scorecard (Expanded). Available at:< http://www.reliabilityweb.com/art07/rcm pmg.pdf>. Accessed on: 01 May 2010, passim.

RELIASOFT Weibull ++7. Reliasoft Corporation.1 CD-Rom. 2009.

SCHWAN, Clair A. Introduction to Reliability Centred Maintenance. **Proceedings of Reliability Centred Maintenance for Substations**, Transmission and Distribution Conference, Electric Utility Consultants, Inc. 1999.

SEIXAS, Eduardo de Santana. **Reliability-Centred Maintenance - Establishing Maintenance Policy Based on Equipment Failure Mechanisms,** 1999. Available at:< http://www.ebah.com.br/manutencao- centred-on-reliability-pdf-a60303.html>. Accessed on: 02 May 2010.

SMITH, Anthony M. **Reliability - Centred Maintenance,** McGraw Hill, 1993

SMITH, Anthony M" HINCHELIFFE, Glenn R. **RCM - Gateway to world class maintenance,** Kindle edition, 2004

Printed by Books on Demand GmbH, Norderstedt / Germany